SOL AND LUNA
The hermetical wedding

STEVEN SCHOOL

Copyright © 2012 Steven School

All rights reserved.

ISBN-10:148266853X
ISBN-13:9781482668537

DEDICATION

I dedicate this book, to those who wish to learn the great work, and who know the meaning of the word integrity.

CONTENTS

	Acknowledgments	i
1	The Rough Stone.	3
2	Sol and Luna.	Pg #5
3	Conjunction.	Pg #11

ACKNOWLEDGMENTS

In this book, we will discuss the rough stone, as well as sol and Luna, their proper conjunction, and the proper type of modern alchemical vessels that are required for completing the great work. We will simplify, the confusion.

1 THE ROUGH STONE.

In this section of my book, we will discuss the rough stone, which is also called by many alchemists, the prima materia, which is also the true starting point of the great work in regards to the philosophers stone, since without this substance, you have nothing.

It has been written, that the ancient philosophers said that the veritable stone was hidden in the cavern of the metals. Some of them also stated in writing that the life of the metals is in their ore, and that the fire of smelting is their death.

Now I ask you, if you were to mend a rubber tire, would you not use another quantity of rubber to mend it?, if you were to mend a broken tree, would you not use the pieces of living wood from the same tree?

If you were to receive a serious burn injury, would not the surgeon remove new skin from your thigh and use it to repair your wound by conducting a skin graft using living tissue?, obviously if dead tissue was used, the result would not be the same.

So if a medicine for metal you would make, then what type of substance would you make this remedy from?, how would you break this substance down?, how would you separate it into its various parts and eliminate its impurity without destroying its vital life force?

Would you use the metal acetate path?, would you use the internet to research knowledge regarding the metal acetate path?

Would you be wise enough to know that alchemical terms are metaphors not to be taken at face value, and would you take the time to see the deeper meaning and thus avoid all toxic substances?

Would you be smart enough to know that since a man cannot be created from a man's hand and a woman's foot, that animal substances such as urine and blood can never be used to create gold?

Would you realize that all harsh acids and toxic chemicals will only destroy the life force of your subject and further serve to pollute your work?

Some substances, which exist on this planet, are toxic by nature, and some of them, when refined, become even more toxic, therefore it is rather obvious that these types of substances would never be used to make a medicine for man, or an elixir of life, substances which take life away, cannot preserve or extend life. If someone has indicated the utilization of radiation, or toxic substances to be used in alchemy, then they are no one's teacher, they do not qualify to be given the title of alchemist, they do not know the true stone, and you must cast their works aside.

These are all great clues, that will save you time and money, and get you on the right track to finding what you seek, at some point in history, myths were based on facts, at one time, something happened which started a rumor and the initial event was important and powerful enough that it became myth and legend, being important enough that it was talked about, and written about for hundreds or thousands of years to follow.

A good majority of people today believe that myths are simply old fairy tales, could it be that we were subconsciously taught to believe this?, could it actually be possible that many myths and legends are actually secret codes that were designed to preserve higher levels of secret information that is known to only a select few?, it is not only correct, it is fact. It is only for those who have eyes to see.

SOL AND LUNA.

If you are studying alchemy, then you probably have heard of sol and luna, you may have seen ancient pictures of the green lion devouring the sun, while the moon is pictured in the water below.

Many people have no clue to what this picture actually means, another pictograph that I found from previous centuries depicts the king and queen having one body with two heads, two arms, and two legs. In that pictograph they are laying on a wooden bed with a great fire underneath, and thick white smoke rising from them, like an ancient funeral pyre. These two pictures that I have described for you, hold more alchemical meaning and knowledge than most people can comprehend. The king and queen are in fact, nothing more, than sol, and Luna.

In Flamel's works, one of his pictures depicts an Athanor and a sand bath, in the sand is an Aludel.

Some of the medieval alchemists used a two piece retort for conducting the great work, the top of which was shaped like a modern glass retort, except theirs was made from copper, and the bottom half of this device was simply an earthen ware or clay crucible. This device was suitable for the great work, but there is a superior way which Flamel and some others understood, it involves the use of a standard retort, followed by the use of an Aludel. We can use a standard borosilicate retort, I prefer the two fifty ml size, and for the Aludel there are many simple ways that we can make this device at home, I will cover the most simplest method, which is also the most effective, and happens to be very similar to what Flamel used for his work.

For a simple homemade Aludel, the best places to gather your materials are flea markets, and yard sales. You can use a vision ware, flame proof glass plate set upon an electric hotplate, place a small borosilicate beaker containing your substance in the center of this plate, now take a glass jar or bowl and invert it, turn it upside down and place it on the glass plate creating a seal.

Sol, is a salt, it is not just any salt, nor is it common sodium chloride or sea salt, however this particular salt is the body of the philosophers stone, it is the white stone, it is known by the symbol of the sun. when Michael Sendivogius wrote about our living gold and our living silver, many people became confused as to which one was which, the white stone is our living silver, yet it is called sol, and known by the symbol of the sun.

Luna, is a certain crystal clear, metallic water. Not ordinary water, it is known by the symbol of the moon, yet it is our living gold, and it is our secret multiplier which is used to make the imbibitions of the white stone, over time, this water will be used to repeatedly dissolve our salt, so that finally the two will fuse together into a new substance and now the king is born. Now the water is called our living gold because over time, it can be made to solidify, once this occurs if it is subjected to gentle coction, (cooking in sand bath also known as the Egyptian fire), it will turn the color of gold, like a golden glass, after which it will turn dark translucent red like a ruby. This liquid will therefore exalt the white stone, into the celestial ruby, and this liquid will also be used for the subsequent multiplications of the red stone. The white stone by itself is already complete, it cannot be multiplied into more white stone, you can either accept it as it is, or bring it to the red stone which can be multiplied in quality and virtue.

Since the salt is the body of the stone, and the liquid provides the color, adding more liquid will darken the stone further adding a more concentrated effect, and thereby increasing its tingeing power. When the two substances are united in the sealed egg and subject to coction in the sand bath on low heat, the two will dissolve and coagulate repeatedly until becoming a solid and fixed substance. Once this point of fixity is reached, leaving the egg undisturbed in the heat will facilitate further color changes, once it becomes red, which comes after gold, then we have completed one turn of the wheel. Our stone must have at least three turns of this wheel before it receives its tingeing power. Some alchemists performed further multiplications on top of this.

Now, back to the antique pictographs which we mentioned earlier, the green lion devouring the sun is actually symbolizing a two part distillation. The ancients mentioned that two separate receptacles were used for this as well as two different heat ranges. What they failed to mention is that two different distillation vessels are used, and in the older times two different furnaces which is the reason for the Athanor and the sand bath which were shown side by side in many old alchemical pictures. In modern times the Athanor is obsolete, we can use an electric hotplate and a crock pot filled with sand. The great work done the first time from scratch will take ten to twelve months to complete in the far superior wet path if you know what you are doing, later on it is possible to advance further into the dry path after you have obtained and properly stored sufficient quantities of sol and Luna. The dry path is very fast, however the finished product does not have nearly the same level of quality and virtue as the stone completed in the wet path. Also the dry path can be very dangerous, the only benefit of it that I can see is that a master alchemist can produce the stone with the dry path, in about an hour. It is done in an iron crucible. The two distillation apparatuses that are used are the borosilicate retort, followed by the Aludel which you must make yourself. Some of the ancients did it with one vessel which was a custom made two piece retort made of copper and clay, but this leads to impurities in your final substances so unless you can get someone to custom make a two piece retort with a borosilicate top, just go with the retort and Aludel method of distillation.

The green lion devouring the sun while the moon watches from the water below, this means that the moon is in the water, which is secretly our living gold, it further means that this substance was distilled over first, at a lower temperature, and that it collected in our receptacle as a clear liquid.

The sun resides higher in the sky, we know that the sun is very hot, this substance is distilled over second, at a much higher heat, and in the Aludel, it forms and solidifies high on the glass. These two substances both come from one root, which means the green lion. Many pictographs show three flowers, or three steps, or three rivers which separate, one river flowing away by itself, while the other two reconnect and become one.

This means that the green lion is a threefold substance, it contains both sol and Luna, and it also contains an impurity which must be separated. In the first half of our distillation we used the borosilicate retort, and thus collected the water, for the second distillation we remove the leftover material from the retort, (the green lion), we place this substance into the beaker in our Aludel, using the electric hotplate for both distillations. Now the heat is turned up high, the matter begins to melt, bubbling that liquid black pitch, then it begins to burn and reduce to cinders, this is the green lion devouring the alchemical sun, sol will now begin to rise as a thick white smoke, when the heat is turned off and the Aludel cools, sol will stick to the glass and solidify as a waxy looking white salt. This is our white stone. It is complete and needs nothing further unless you will exalt it to the red, by doing the imbibitions with the water which was brought over first.

Clear your mind of all the false recipes and ingredients you have seen spread all over the internet. With the series of short books that I have written you have plenty of ability to complete the great work provided that you contribute a little thought and willpower, the most oppressive issue which plagues people the most is confusion, the work is simple, but by looking at all the false recipes and false ingredients rumored worldwide, it is easy to become very confused and this leads to one's downfall, which is referred to as being lost in the labyrinth, I went through it myself, and it is very frustrating. My books are meant to keep it simple and help you avoid the labyrinth.

By now you should have a pretty good idea of sol and Luna, the two ingredients of the red stone, the white stone has only one ingredient, sol. The red stone has a considerable weight to it, this is because liquid is heavy, when the salt and liquid coagulate into a crystalline glass like substance, it has become a solid salt, but it retained the weight of the liquid that became salt.

The pictograph of the king and queen on the burning funeral pyre with the white smoke. This indicates that the two are one, sun and moon, sol and Luna, it also signifies that heat is used, up to the highest degree or equivalent of open flame, it also shows that the two vaporize and rise up into the air. It means that they are one thing, that separates during distillation, but will be rejoined during conjunction. Or the chemical/hermetical wedding.

There is an old mathematical equation associated with the philosophers stone, from one is made three, from the three two, and from the two, one.

This means that from the green lion we separate sol, Luna, and the impurity which is removed in the Aludel and cast aside, sol and Luna are then reunited into one substance (the hermetical wedding), and once three turns of the wheel are completed the king is born, our stone of the third power or multiplication.

If the stone is to be used for medication, you must remember that it is very powerful, the proper dosages are known to the masters, others should not even delve into that since a simple miscalculation in either preparation, or dose, can cause fatal results. The further the stone is multiplied, the further it must be diluted for medicinal use as its potency becomes ten times stronger with each turn of the hermetical wheel.

I like to work on my hobby of alchemy year round, but all winter long, I simply cannot wait for summer to arrive, when I can venture outdoors, working outside amongst the trees, the plants and the animals underneath the great blue sky, the solar power of the sun, which is the center of our universe, is the perfect thing for completing many aspects of the great work, I have also heard that god resides in the center of our universe, and that Jesus is the supernatural son of the sun. the son of god, and god in the flesh. It could be that part of Gods work is to light and heat the earth, without which we would not be.

Vitriol, many modern students have wondered what this substance was to the ancients, I will fully define it for you now, it is an abbreviation of the Latin phrase Visita Interior Terrae Rectificando Invenies Occultum Lapidem.

In modern terms it means visit the interior of the earth, and there by rectification find the lapis. Our hidden stone, the earth they are referring to is the green lion, by rectification is meant distillation.

I have spent years pondering the writings of Michael Sendivogius, in particular the new chemical light.

From my notes of interpreting his works I will give you this little gem, "our air", the white smoke condensed is the womb, this is the receptacle of "our water", which is the mineral seed. If you are an adept and familiar with Michaels work, then you can understand this, for the rest of you, I have plainly laid it down in simple terms, it took me five years to understand his work. As I unfolded the mystery on my own, the writings of ancient philosophers became clearer to me since with each of my own alchemical breakthroughs, I could further see and understand what the old ones were doing in their work. Knowledge is a gift.

I do not knock the work of other aspiring alchemists, it is a path that we must traverse in order to find the truth. Certain stepping stones must be traveled to attain mastery over this subject. We are all of the same art, seeking to obtain the same goal. Those who have anger in their hearts

towards another alchemist, are simply jealous of his work. That is not the way to go through life, anger and jealousy will only obstruct the students vision, rendering he or she, unable to see the light of truth. It does not matter who reaches the goal first, just like any other endeavor in life, you get back what you put into it. If you will see the finish of the great work, then you must put forth the required effort, just as I did, however with my work I have simplified many of the obstacles for you, lighting the way so to speak. As my knowledge, understanding, and experience grows, I write new books, further illuminating the secrets. I wish that books like this were available for me to study when I was coming up, but there were none, there was only the blind confusion and frustration of the labyrinth, in which most people become hopelessly lost and never find the way through, perseverance is what it took, plain determination and a refusal to give up.

SOL AND LUNA.

CONJUNCTION.

In the works of Sendivogius, who was a disciple of the accomplished alchemist Alexander Seton, he mentions three principles, our living gold, our living silver, and our earth, he also lists somewhat obscurely, his recipe for confecting the stone in the wet path of ten months, his recipe calls for only one imbibition, which is due to the fact that he had a method of tripling the multiplier in the first imbibition. We will cover this in this section of my book on conjunction.

Let us uncover a secret portion of his work, we will start with his three ingredients which are actually only two, it is a different method of achieving the same outcome, and it is a hybrid cross of part wet path, part dry path.

As we mentioned earlier, what he meant by our living silver is sol, the white salt. What he referred to as our living gold, is actually Luna, the living water or clear distillate. He had developed his own technique, of gently evaporating the crystalline water to dryness in a sand bath on very low heat, this left him with a clear salt which reflects a myriad of colors in the light, this substance at this stage is referred to as the peacocks tail. Michael called it, our earth. He used a glass globe, in modern times it is known as a round bottom flask, his had a long neck of about six inches in length, and utilized a ground glass stopper.

In this philosophical egg he placed eleven parts of our earth, two parts of our living silver, and one part of our living gold. He sealed the glass and subjected this egg to the heat of our central fire for ten months. This secret central fire means that he used no external heat for this portion of the work, because when sol and Luna are combined an endothermic reaction takes place, the two substances combined will create their own heat from within. In his day, alchemists would bury this egg in the dirt down in the basement for roughly anywhere from nine to twelve months, they would cover the neck of the vessel with two hollowed out oak rounds to protect it from cold. The alchemist could lift the wood at any time to look inside his glass without disturbing the egg.

In ten months of time or roughly there about, the water element, our living gold, will dissolve the two salts, it will eat through the elements, and then coagulate them together into the glass like substance of a red color, which when ground to powder will appear as a lemon yellow colored crystalline salt, which is the famous transmutation powder of the medieval alchemists. Note- variables in the heat range used and number of multiplications can result in a darker powder sometimes red, or even black, the darker it is, the more powerful it is.

When Luna is reduced to its dry form, it can also be used to perform the dry path, Sol is already a dry waxy salt, if you have both of these two substances prepared in their dry form, they can be gently melted together in the crucible and fused into the red stone, however as I said before, the dry path is very dangerous, the wet path is safe, so let us go with safety, and a more virtuous stone.

Now the term multiplication in regards to alchemy and the philosophers stone seems to be a stumbling block which confuses many people, even some who consider themselves to be adepts, to this I say the mark of an adept is one who has made the stone. If you saw my book alchemy survival guide on amazon.com then you know my work because it has a picture of my red stone right on the cover, my work speaks for itself. I will simplify the art of multiplication for you. Many words have been used to describe one thing, conjunction, imbibition, multiplication, it is all the same thing, it simply means to pour the water on the salt, then let your sealed egg be subject to coction, nature will dissolve and coagulate and your slurry will become a fixed stone. The first time this is completed you have the stone of the first order, which means you have multiplied it one time, external heat will now bring it through the colors, and at ruby red, it is ready for another imbibition or multiplication.

Every alchemist has his or her own special way of confecting the lapis, this accounts for various differences in the works of one versus another, it all leads to the same goal, but there is more than one way to skin a cat. I do not feel that one the method of one person is superior to another, I simply say to each his own, we are all on the same path to the same goal, it matters not, who gets there first, Paracelsus referred to the water of life as oil, he liked to bake his egg in his Athanor after conjunction, this is also fine, it greatly reduced the time since with his method each multiplication could be completed in twenty days. Whether there were variances in the quality of Sendivogius stone as compared to Paracelsus, is unknown to me, adding external heat prematurely can speed the work, however if you are not careful, it can also ruin your work.

SOL AND LUNA.

Let us now look at the way in which some modern alchemists undertake the task of conjunction in the wet path, some of the differences between the old ways and the new methods are because we now have modern equipment and electricity, and also the fact that we like to experiment and advance with new technical breakthroughs. Once we have prepared our living gold, and our living silver, which is to say Sol, our white salt, and Luna, our water which is also the alkahest, we shall grind the salt to powder in a clear glass mortar and pestle, and then place it in a round bottom borosilicate flask. The flask will have plenty of empty space leftover to allow for expansion, now I use a clear glass bottle with a built in eyedropper for the imbibitions, simply drip some of your clear distillate into the white salt until you have applied only just as much of the liquid as the salt can absorb and no more, if you accidently put in too much liquid do not worry, it will still work, it will just take a little longer.

Now seal your egg with its ground glass stopper, and place it in your sand bath. I use an electric crock pot from the grocery store, filled with play sand from home depot for this. Most crock pots have three temperature settings, low, medium, and high. Mine has four settings, warm, low, medium, and high. Bury your egg in the sand so that it is secure from falling over, but leave the neck of the glass exposed so that you can periodically peer inside and check the progress of your work without disturbing your flask.

Choose the lowest heat setting, your substance will soon begin to undergo changes, it will dissolve and coagulate, becoming semi fixed, and then dissolve again, it will undergo slight color changes, you may see a mostly white substance, it might start to have a greenish tint, it will show signs of golden or citrine color, leave it alone until it becomes a completely fixed and solid substance, or very close to that, and now raise the heat, turn it up to medium for a few days, and then on high until it turns red and becomes the celestial ruby.

At this point unplug your sand bath and let it cool. You have completed one turn of the alchemical wheel, and it is at this point that you would remove half of your stone to keep for medicinal purposes. Store it in a sealed glass vial for future use. Now the remainder of the substance which is left in the philosophical egg, we will reimbibe it just as we did the first time and repeat the entire procedure again. To give our stone ingress and power over the metals we must have a minimum of three turns of this wheel, unless you are doing the dry path or the Sendivogius method.

Most alchemists liked to perform four turns of the wheel or multiplications, some went all the way up to the seventh power. I have never heard of anyone going farther than this, nor do I know what the outcome would be, however I do know that the stone of the fourth order is plenty powerful enough, since each multiplication increases its tingeing power by tenfold.

Now the rumor of the secret alkahest being able to dissolve gold like ice in warm water is false which I have learned on my own, through my work and experiments. The truth is that it is indeed just another metaphor which the old ones loved to use. The truth of it is that the clear distillate, which is Luna, is the Alkahest, and if you drop sol into this water, the salt will dissolve like ice in warm water, which was Gualdus method for confecting the stone. He liked to call Sol sea salt to add to the confusion. Others referred to Luna with metaphors such as rain water, May dew, urine, the menstrual blood of the whore, mercury, our living pontic water, and other such covert code names.

Now the type of method that Gualdus used is the same technique of Hermes Trismegistus, Sol, the white salt, is the bird of Hermes. The whites doves of Diana that you may have seen a picture of a flask partially filled with liquid, and small white doves diving into the liquid.

The way this method worked is that the alchemists put Luna, the clear distillate into the flask, with a large empty space left over. The ratio of liquid to salt will be not be exact, however it will be mostly liquid so that this method will only require one turn of the wheel to complete, so place your round bottom flask containing the correct liquid securely into a flame proof pan filled with sand, set it on your electric hotplate, (Gualdus used coals), now grind your salt (Sol), in a clear glass mortar and pestle to powder, begin by dropping a small amount of the salt into the flask and insert the ground glass stopper. Check it daily and continue adding more of the doves of Diana a little at a time until the liquid begins to thicken, now leave the sealed egg alone in the gentle heat of the sand bath, it is going to take a matter of months, do not disturb it, let it be subject to coction, (cooking), until it is completely fixed and has gone through the colors finally resulting in a dark, ruby red, glass like substance, at which time, this work is completed.

SOL AND LUNA.

I recommend that if you are going to undertake the great work of alchemy, that you begin collecting an assortment of mortar and pestles. I have a molcajete, which is a large mortar and pestle made from volcanic rock, some hispanic people traditionally use these to make guacamole, they are perfect for plant alchemy so that you can grind your dried plant material.

In regards to preparing the prima material for the magnum opus, a cast iron mortar and pestle works best, you could also use brass or stainless steel, however when grinding Sol, or Luna in its dry form, we need to use clear glass for this.

I also have an assortment of marble and soapstone mortars for grinding calcined ashes, this is good for making an animal stone, or a plant stone.

The volcanic rock molcajete can be used for grinding the prima material in the magnum opus, but metal is superior in this department.

My first Aludel which I made myself was using a glass globe from a kerosene lamp, it very quickly afforded me a quantity of the white stone, however a sealed system is much more efficient, using an inverted bowl or jar instead works better, coupled with the two part distillation that I previously mentioned earlier in this book. The jar or bowl, whichever you decide to use, should be narrower at the opening, than the flame proof glass plate that it rests upon, thus creating a seal. The jar will work better than the bowl and I would recommend using a taller one such as a tall narrow pickle jar because the more open interior space of clean glass that you have available, the more purified white stone you will collect with each distillation.

Redistill the same ashes a few times to insure that you have collected all of the salt.

Sol is called fixed, and Luna is called volatile, this is because Sol is a solid, dry salt, however they are both volatile, it just takes a much higher degree of heat to volatize sol, whereas Luna will volatize in low heat.

Even the fixed red stone will volatize in a very high degree of heat.

Most alchemists when conducting the great work, will separate out half of their white stone, storing it in a sealed glass such as a spice jar, the other half they will exalt into the red stone of the first order.

They will then separate out half of this red stone of the first order and store it in sealed glass, these are kept for future medicinal use, or to later multiply and exalt. The other half of the red stone of the first order is left inside the philosophical egg and multiplied to at least the third order or higher, to be used as a medicine for metals. If you have any leftover Luna from any of your works, simply place it in an open spice jar and on very low heat in sand bath let it evaporate to dryness, take it out of the gentle warmth just before it is completely dry so that the color changes do not begin, when it finishes drying on its own it is now in the stage of the peacocks tail, seal the jar and store it for future use. I utilize a wooden wall mounted spice rack and clean, dry spice jars for storing these substances.

SOL AND LUNA.

I have written several other books on alchemy and other subjects, if you have additional areas of the work that you are unsure of, my other titles should clear up the matter for you. All of my books are available on amazon.com, and they include-

Alchemy and the golden water.
Alchemy and the green lion.
Alchemy and the peacocks tail.
Alchemy survival guide.
How to make money.
Trophy wife.
Grandmas delicious recipes.
Karate secrets revealed.
Casino survival guide.
Wilderness survival tips.

I sincerely hope that you have enjoyed my book, good luck in your work, and remember, if you would like to see a picture of my red stone, go to amazon.com and look up alchemy survival guide, the picture of my stone is on the cover, I may in the future write a new book called Alchemy and the white stone, I will place a picture of the finished white stone on that book cover.

Good day to you.
Steven School.

Made in United States
Orlando, FL
04 April 2025